CHEMISTRY QUIZ

Preface

This book is genuinely written for revising the fundamental concept of chemistry. It is aimed to the secondary level students. It can serve as a reference for a particular topic. It is also useful for various competitions.

Index

This book can be treated as a general chemistry quiz book for secondary level students. It includes all fundamental chapters of chemistry. For convenience of study it is divided into two parts.

Part 1 includes chapter pertaining to some elementary concepts of ch3emistry, structure of atom, atom and molecules and matter around us. Part 2 includes chemical reactions and equations, acid, bases and salt, metals and non metals, carbon compounds and the last lesson deals with periodic classification of elements.

CHAPTER 0

SOME ELEMENTARY CONCEPTS

SUPPLEMENTRY PROBLEME

MCQ

Q1. Nucleus consists of?

a) Electron and neutron

b) Electron and proton

c) Proton and neutron

d) None of these

Q2. Valency of oxygen is?

a) 2

b) -2

c) 1

d) None of these

Q3. $Na + Cl_2 \rightarrow$?

What is the product of above reaction?

a) $NaCl_2$

b) $NaCl_3$

c) $NaCl$

d) None of these

Q4. How many elements has synthesized in lab till now?

a) 22

b) 26

c) 92

d) 118

Q5. Which is an inert gas?

a) Hydrogen

b) Nitrogen

c) Oxygen

d) Neon

Q6. Which one is the heaviest among these?

a) Electron

b) Proton

c) Neutron

d) All has same masses

Q7. Which one has independent existence in a chemical reaction?

a) Electron

b) Atom

c) Molecule

d) None of these

Q8. How many electrons are there in a Na+ ion?

a) 10

b) 11

c) 12

d) 9

Q9. Why does any atom react with other?

a) to give its electron

b) to gain electrons

c) to get a stable configuration.

d) none of these

Q10. Cl_2 is ?

a) Monatomic

b) diatomic

c) Inert gas

d) none of these

Choose whether these statements are true or false

a) Electrons have negative charge.

b) In a simple hydrogen atom one electron and one proton is present.

c) Electrons revolve around the nucleus.

d) Electronic configuration is helpful in finding whether a given element is inert or not.

e) Chlorine has positive valency.

f) Hydrogen molecule is diatomic.

CHAPTER 1

STRUCTURE OF ATOM

SUPPLEMENTRY PROBLEME

MCQ

Q1. Cathode ray consists of?

a) Electron

b) Atom

c) Molecule

 d) None of these

Q2.canal ray consists of?

a) Electron

b) Atom

c) Molecule

d) None of these

Q3. Neutron has …………..charges?

a) Negative

b) Positive

c) None of these

d) Some time negative some time positive

Q4. When cathode ray observed in J.J Thomson's experiment?

a) At high temperature and high pressure.

b) At very high voltage and very low pressure.

c) At high voltage and high pressure.

d) None of these

Q5. Who discovered electron?

a) J.J Thomson's

b) E. Goldstein

c) James Chadwick

d) Neils bohr

Q6. . Who discovered neutron?

a) Rutherford

b) E. Goldstein

c) James Chadwick

d) Neils bohr

Q7. Isotopes have same……………?

a) Atomic masses

b) no. of neutrons

c) Atomic number

d) none of these

Q8.Cl-35/17 and Cl-37/17 are?

a) Isotopes

b) Isobars

c) Isotones

d) None of these

Q9. N-14/7 and C-14/6 are?

a) Isotopes

b) Isobars

c) Isotones

d) None of these

Q10. In alpha particle scattering experiment Rutherford chosen?

a) Zn foils

b) Silver foil

c) Gold foil

 d) None of these

State whether these statements are true or false

a) Rutherford gave the plum pudding model of atom.

b) Somerfield explained that the electrons revolve in an elliptical orbit.

c) Isotones have same no. of protons.

d) The whole mass of atom is concentrated in the nucleus.

e) James Chadwick discovered the neutron.

f) Nucleus is neutral.

g) Atom is neutral but nucleus is negatively charged.

h) Isotope of iodine is used to cure goiter.

CHAPTER 2

ATOM AND MOLECULES

SUPPLEMENTRY PROBLEME

Q1. The word atom was given by?

a) Bohr

b) Rutherford

c) Lavoisier

d) Dalton

Q2. Law of conservation of mass is given by?

a) Bohr

b) Rutherford

c) Lavoisier

d) Dalton

Q3. Molecular mass of $CuSO_4$ is given what? Given (Cu=63u, S=32u, O=16u)

a) 157u

b) 159g

c) 159u

d) 1.59u

Q4. The number of entities in one mole is equal to?

a) 1.6×10^{-19}

b) 1.6×10^{-27}

c) 6.022×10^{23}

d) None of these

Q5. Total number of electrons present in 44g of CO_2 gas is?

a) 1 mol

b) 11 mol

c) 22mol

d) None of these

Q6. When 196g of HCl is dissolved in 5 litre water. What is the molarity of the solution?

a) 0.2 mol/litre

b) 0.4 mol/litre

c) 2 mol/litre

d) None of these

Q7. How many sodium ions (Na+) are there in a 58gram of NaCl? (given Na=23u, Cl=35u)

a) 2 mol/litre

 b) 1 mol/litre

c) 0.5 mol/litre

d) none of these

CHAPTER 3

MATTER

SUPPLEMENTRY PROBLEME

MCQ

Q1. Which has fixed shape and volume?

a) Solid

b) Liquid

c) Gas

d) Plasma

Q2. In which intermolecular space is Maximum?

a) Solid

b) Liquid

c) Gas

d) all has same

Q3.difussion is involved in?

a) Solid

b) Liquid

c) Gas

d) None of these

Q4. The process of changing solid to gas directly is known as?

a) Evaporation

b) Condensation

c) Sublimation

d) Melting

Q5. On adding heat the intermolecular forces between two molecules:

a) Increases

b) Decreases

c) First increases then decreases

d) None of these

CHAPTER 4

PURE AND IMPURE MATTER

SUPPLEMENTRY PROBLEME

MCQ

Q1. NaCl is a?

a) Pure substance

b) Impure substance

c) Mixture of pure and impure

d) None of these

Q2.solution of sugar in water is?

a) Element

b) Compound

c) Homogeneous mixture

d) Heterogeneous mixture

Q3. Milk is a?

a) Colloidal

b) Solution

c) Suspension

d) Compound

Q4. Cream is obtained from milk using which processes.

a) Filtration

b) Chromatography

c) Sublimation

d) Centrifugation

Q5. Which process is used to separate drugs from blood?

a) Filtration

b) Chromatography

c) Sublimation

d) Centrifugation

Q6. Which process is used to squeezes water from cloth in washing machine?

a) Filtration

b) Chromatography

c) Sublimation

d) Centrifugation

CHAPTER 5

CHEMICAL REACTIONS AND EQUATIONS

Review Questions

Q1. When Zinc reacts with HCl which gas is released?

a. H_2

b. He

c. O

d. Cl

Q2. In equation the given reaction, which is reactant?

$H_2 + O2 \rightarrow H2O$

a. H_2

b. O2

c. Both

d. None

Q3. $Fe + H_2O \rightarrow ? + H_2$

Which compound should be at the place of ? in the above equation?

a. Fe_2O_3

b. FeO

c. Fe_3O_4

d. $Fe(OH)_3$

Q.4 $3Fe + 4H_2O \rightarrow ? + 4H_2$

Which compound should be at the place of ? in the above equation?

a. Fe_2O_3

b. FeO

c. Fe_3O_4

d. $Fe(OH)_3$

Q5. $C + O_2 \rightarrow CO_2$ is ?

a. Displacement reaction

b. Endothermic reaction

c. Exothermic reaction

d. None

Q6. What is the formula of lime stone?

    a. CaO

    b. $CaCO_2$

    c. $Ca(OH)_2$

    d. $CaCO_3$

Q7. what is the formula of quick lime?

    a. CaO

    b. CaCo2

    c. Ca(OH)2

    d. CaCO3

Q.8 what should be the value of X so that equation becomes balanced?

$2Pb(NO3)2(s) \rightarrow 2PbO(s) + XNO2(g) + O2(g)$

    a. 1

    b. 2

    c. 3

    d. 4

Q.9 $Fe(s) + CuSO4(aq) \rightarrow FeSO4(aq) + Cu(s)$

Which type of reaction is this?

    a. Displacement reaction

    b. combination reaction

    c. decomposition reaction

    d. None

Q.10 which is oxidised in the given reaction?

$2Cu + O_2 \rightarrow CuO$

    a. Cu

    b. O2

c. CuO

d. None

Q.11 which is essential for corrosion?

a. air

b. water

c. moisture

d. all of these

Q.12 silver articles become black when exposed to air. In this process which compound is formed?

a. Silver oxide

b. Silver nitrate

c. Silver sulphide

d. Silver carbonate

Q13. Food kept in air tight jar prevent it from?

a. oxidation

b. reduction

c. corrosion

d. None

Q14. Chips packets are flushed with?

a. N

b. O_2

c. Cl_2

d. He

Q.15 $2Cu + O_2 \rightarrow 2CuO$

Which type of reaction is this?

a. Displacement reaction

b. combination reaction

c. decomposition reaction

d. None

Q16. In which type of chemical reaction ions are interchanged?

a. Displacement reaction

b. combination reaction

c. decomposition reaction

d. double displpacement reaction

Q17. Opposite of combination reaction is?

a. Displacement reaction

b. combination reaction

c. decomposition reaction

d. double displpacement reaction

Q18. Oxidation is the gain of?

a. oxygen

b. electron

c. proton

d. none of these

Q19. Reduction is the gain of?

a. oxygen

b. electron

c. proton

d. hydrogen

Q20. Precipitates are?

a. Soluble in water

b. Insolu ble in water

c. Partially soluble

d. Float in water

Math the followings

Q21.

A	B
AgNO$_3$	combination reaction
2H$_2$ + O$_2$ → 2H$_2$ +O$_2$	black
rancidity	unbalanced reaction
Fe$_2$O$_3$ + Al→Al$_2$O$_3$ + Fe	nitrogen

Q22.

A	B
oxidation	gain of electron
reduction	loss of electron
corrosion	chips
rancidity	iron

Q23.

A	B
Black and white photography	thermite reaction
Exothermic reaction	AgBr
CuSO$_4$	Au
Noble metal	blue

CHAPTER 6

Acids, Bases and Salts

Review questions

Q1. Acid changes blue litmus to?

    a. red

    b. pink

    c. yellow

    d. no change

Q2. Phenolphthalein and methyl oranges are

    a. synthetic indicator

    b. natural indictor

    c. both

    d. none

Q3. In olfactory indicators?

    a. Colour changes

    b. Taste changes

    c. Odour changes

    d. none

Q4. Acid on reaction with metal produces?

    a. H_2

    b. O_2

    c. N_2

    d. None

Q5. Acid on reaction with metal hydrogen carbonate produces?

    a. H_2

    b. CO_2

    c. N_2

    d. None

Q6. In aqueous solution acid produces which type of ions?

    a. H^+

    b. OH^-

    c. N_2

    d. None

Q7. The process of dissolving an acid or a base in water is highly?

    a. endothermic

    b. exothermic

    c. cold

    d. None

Q8. pH scale measure which ion in a solution?

    a. H^+

    b. OH^-

    c. Cl^-

    d. None

Q9. A solution has pH value 7. It is likely to be

    a. acid

    b. base

    c. salt

    d. none of these

Q10. Our body works within pH range?

    a. 2-4

    b. 12-14

    c. 6-10

    d. 7-7.8

Q11. Milk of magnesia is?

    a. acid

b. weak acid

c. antacid

d. None

Q12. Acetic acid is present in?

a. Apple

b. guava

c. vinegar

d. None

Q13. Oxalic acid is present in?

a. potato

b. tomato

c. onion

d. none

Q14. pH value of a salt is?

a. 6

b. 7

c. 8

d. 9

Q15. Formula for bleaching powder is?

a. $CaOCl_2$

b. $CaOCl$

c. $CaCl_3$

d. None

Q16. Na_2CO_3 is?

a. Washing soda

b. Sodium carbonate

c. Sodium hydrogen carbonate

d. None

Q17. Which is used for removing permanent hardness of water?

    a. $CaSO_4.1/2H_2O$
    b. $CaSO_4.2H_2O$
    c. $CaSO_4$
    d. Washing soda

Q18. A solution turns blue litmus red its pH is likely to be?

    a. 1
    b. 7
    c. 8
    d. 9

Q19. A solution reacts with crushed egg-shells to give a gas that turns lime-water milky. The solution contains
(a) NaCl

(b) HCl

(c) LiCl

(d) KCl

Q20. Which one of the following types of medicines is used for treating indigestion?
(a) Antibiotic
(b) Analgesic
(c) Antacid
(d) Antiseptic

Match the following:

Q21.

        A   B

    Acetic acid   nettle sting

Citric acid	curd
Lactic acid	vinegar
Methanoic acid	orange

Q22.

A	B
Bleaching powder	$CaSO_4.2H_2O$
Baking soda	$CaSO_4.1/2H_2O$
Washing soda	$CaOCl_2$
Plasterof paris	$Na_2CO_3.10H_2O$
Gypsum	$NaHCO_3$

Q23.

A	B
acid	red cabbage
base	salt
pH 7	H^+
natural indicator	OH^-

Chapter 7

Metals and Non Metals

Review questions

Objective questions:

Q1. Which metal is liquid at room temperature?

    a. Hg
    b. Br
    c. Cl
    d. None

Q2. Which non metal is liquid at room temperature?

    a. Hg
    b. Br
    c. Cl
    d. None

Q3. Which non metal is good conductor of electricity?

    e. Cu
    f. Fe
    g. Graphite
    h. Diamond

Q4. Which of the following can be cut by knife?

    a. Na
    b. Mg
    c. K
    d. All of the above

Q5. Calcination is used for?

    a. Sulphide ore
    b. Carbonate ore
    c. Oxide ore
    d. None of these

Q6. Which one is more reactive?

a. Cu
b. Pb
c. Fe
d. Zn

Q7. In thermite reaction the product is?

a. Fe(l)
b. Al_2O_3
c. heat
d. All of the above

Q8. In galvanization the layer of which metal is coated on iron or steel articles?

a. Zn
b. Mg
c. K
d. none of these

Q9. An alloy is a homogeneous mixture of?

a. Two or more metals
b. Metal and non metals
c. Both a and b
d. None of these

Q10. Al_2O_3 is?

a. Metallic oxide
b. Non metallic oxides
c. Amphoteric oxide
d. None

Q11. Which types of substances are effective in cleaning the vessels?

a. sour
b. oily
c. sweets

d. none of these

Q12. In electrolytic refining the pure metal is taken as?

a. anode
b. cathode
c. anyone can be taken
d. none of these

Q13. Copper is used to make hot water tank because?

a. It is good conductor of heat
b. It is good conductor of electricity
c. malleability
d. ductility

Q14. Coins are manufactured from silver and copper because?

a. These are good conductor of heat
b. These are good conductor of electricity
c. Malleability
d. ductility

Q15. Non metallic oxides are?

a. acidic
b. neutral
c. either acidic or neutral
d. none of these

Q16. Alloy of Cu and Zn is?

a. bronze
b. solder
c. brass
d. amalgam

Q17. In amalgam which element must be present?

a. Hg

b. Pb

c. Fe

d. Zn

Q18.solder has?

    a. High melting point

    b. Low melting point

    c. Moderate melting point

    d. Extremely low melting point

Q19.cinebar is ore of?

    a. Hg

    b. Pb

    c. Fe

    d. None of these

Q20. Copper pyrite is ore of?

    a. Cu

    b. Pb

    c. Fe

    d. Zn

Match the following:

Q21.

A	B
amphoteric	CO_2
acidic	Al_2O_3
basic	NaOH

neutral	H_2O

Q22.

A	B
ionic	CH_4
co-valent	NaCl
roasting	$ZnCO_3$
calcination	ZnS

Q23.

A	B
solder	Zn & Cu
brass	Cu & Sn
bronze	Hg & Al
amalgam	Pb & Sn

Chapter 8

Carbon and its compounds

Review questions

OBJECTIVE QUESTIONS:

Q1. The % of carbon in earth crust is?

   a. 0.002
   b. 0.02
   c. 20
   d. 0.2

Q2. Carbon has valency?

    a. 1

    b. 2

    c. 3

    d. 4

Q3. Catenation is found in?

    a. C

    b. O

    c. F

    d. None of these

Q4. Bonding in carbon is?

a. covalent

 b. ionic

c.co-ordinate

d. none of these

Q5. A saturated hydrocarbon has?

    a. Single bond

    b. Double bond

    c. Triple bond

    d. None of these

Q6. Formula of benzene is?

    a. C_2H_6

    b. C_6H_6

    c. C_5H_6

    d. None of these

Q7. Homologous series differ by?

a. CH_2 unit

b. 14 amu by mass

c. Both a and b

d. None of these

Q8. $CH_3CH_2CH_2OH$ is?

    a. alcohol

    b. ketone

    c. carboxylic acid

    d. None of these

Q9. In oxidation reaction the oxidising agent is?

    a. Alkaline $KMnO_4$

    b. Acidified $K_2Cr_2O_7$

    c. Both a and b

    d. None of these

Q10. In addition reaction the catalyst used is?

    a. H_2

    b. Acidified $K_2Cr_2O_7$

    c. nickel

    d. None of these

Q11. The conversion of ethanol to ethanoic acid is?

    a. Oxidation reaction

    b. Addition reaction

    c. Substitution reaction

    d. None of these

Q12. Which is commonly known as alcohol?

    a. ethanol

    b. ethanoic acid

    c. methanol

d. None of these

Q13. Vinegar is formed from?

   a. ethanol
   b. ethanoic acid
   c. methanol
   d. None of these

Q14. In esterification the catalyst used is?

   a. base
   b. acid
   c. salt
   d. None of these

Q15. Saponification is the reverse process of?

   a. carbocation
   b. esterification
   c. carbocation
   d. None of these

Q16. Ionic end of soap dissolves in?

   a. water
   b. oil
   c. acid
   d. base

Q17. Shampoos are?

   a. soap
   b. detergent
   c. salts
   d. acids

Q.18 which of the following can go under addition reaction?

a. C_2H_6

b. C_6H_6

c. C_5H_6

d. None of these

Q19. CH_3COOCH_3 is?

a. soap

b. acid

c. base

d. ester

Q20. Which of the following causes hardness of water?

a. Calcium salt

b. Magnesium salt

c. Both a and b

d. None of these

Match the following:

Q21.

A	B
C-60	saturated
ethene	allotrops
methane	unsaturated
carbon	catenation

Q22.

A	B
-OH	carboxyllic acid
-CHO	ketone

-CO-	aldehyde
-COOH	alcohol

Q23.

A	B
nickel	substitution reaction
sun light	addition reaction
alkaline $KMnO_4$	oxidation reaction
hot conc. H_2SO_4	formation of ethene

Chapter 9

Periodic classification of elements

Review questions

Objective questions:

Q1. Doberiener arranged the elements in the form of?

a. triads
b. octaves
c. tetrads
d. none

Q2. Newlands arranged how many elements?

a. 56
b. 63
c. 114
d. 118

Q3. Mendeleev left gape for which element in his periodic table?

a. scandium

b. gallium

c. germanium

d. all of these

Q4. Mendeleev used basic concept of?

a. Physical properties

b. Chemical properties

c. Atomic masses

d. all of these

Q5. Who prepared modern periodic table?

a. newland

b. doberiener

c. mendeleev

d. moselley

Q6. How many periods are there in modern periodic table?

a. 18

b. 7

c. 20

d. None of these

Q7. What is the valency of magnesium?

a. 1

b. 2

c. 3

d. 4

Q8. How many groups are there in modern periodic table?

a. 18

b. 7

c. 20

d. None of these

Q9. In periods going right from left, the valency?

   a. increases
   b. decreases
   c. first increases then decreases
   d. None of these

Q10. In groups going from top to bottom, the valency?

   a. increases
   b. decreases
   c. first increases then decreases
   d. remain same.

Q11. In periods going right from left, atomic size?

   a. increases
   b. decreases
   c. first increases then decreases
   d. None of these

Q12. In groups going from top to bottom, the atomic size?

   a. increases
   b. decreases
   c. first increases then decreases
   d. remain same.

Q13. In groups going from top to bottom, the metallic properties?

   a. increases
   b. decreases
   c. first decreases then increases
   d. remain same.

Q14. In periods going right from left, the non metallic properties?

a. increases
b. decreases
c. first increases then decreases
d. None of these

Match the followings

Q15.

A	B
Law of octaves	moselley
triads	mendeleev
gaps in table	newlands
place for isotopes	doberiener

Q16.

A	B
56 elements	moselley
63 elements	mendeleev
9 elements	newlands
118 elements	doberiener

ANSWER TO THE SUPPLEMENTRY PROBLEMS

CHAPTER 0

1.C 2.B 3.C 4.C 5.D 6.C 7.C 8.A 9.C 10.B

TRUE   A, B, C, D, F,

FALSE   E

CHAPTER 1

1.A 2.D 3.C 4.B 5.A 6.C 7.C 8.A 9.B 10.C

TRUE   B, D, E, H

FALSE   A, C, F, G

CHAPTER 2

1.D 2.C 3.C 4.C 5.C 6.B 7.B

CHAPTER 3

1.A 2.C 3.C 4.C 5.B

CHAPTER 4

1.A 2.C 3.A 4.D 5.B 6.D

Answers to the review questions

Chapter 5

Chemical Reactions and Chemical Equations

Q1.a

Q2.c

Q3.a

Q4.c

Q5.c

Q6.a

Q7.c

Q8.b

Q9.a

Q10.a

Q11.d

Q12.b

Q13.a

Q14.a

Q15.b

Q16.d

Q17.c

Q18.a

Q19.b

Q20.b

Q21.

a---b

b---a

c---d

d---c

Q22.

a---b

b---a

c---d

d---c

Q23.

a---b

b---a

c---d

d---c

Chapter 6

Acid, Bases and Salts

Q1.a

Q2.a

Q3.c

Q4.a

Q5.b

Q6.a

Q7.b

Q8.a

Q9.c

Q10.d

Q11.c

Q12.c

Q13.b

Q14.b

Q15.a

Q16.b

Q17.d

Q18.a

Q19.b

Q20.c

Q21.

a---c

b---d

c---b

d---a

Q22.

a---c

b---e

c---d

d---b

e---a

Q23.

a---c

b---d

c---b

d---a

Chapter 7

Metals and Non Metals

Q1.a

Q2.b

Q3.c

Q4.d

Q5.b

Q6.b

Q7.d

Q8.a

Q9.c

Q10.c

Q11.a

Q12.b

Q13.a

Q14.c

Q15.c

Q16.c

Q17.a

Q18.b

Q19.a

Q20.c

Q21.

a---b

b---a

c---c

d---d

Q22.

a---b

b---a

c---d

d---c

Q23.

a---d

b---a

c---b

d---c

Chapter 8

Carbon Compounds

Q1.b

Q2.d

Q3.a

Q4.a

Q5.a

Q6.b

Q7.c

Q8.a

Q9.c

Q10.c

Q11.a

Q12.a

Q13.b

Q14.b

Q15.b

Q16.a

Q17.b

Q18.a

Q19.d

Q20.c

Q21.

a---b

b---c

c---a

d---d

Q22.

a---d

b---c

c---b

d---a

Q23.

a---b

b---a

c---c

d---d

Chapter 9

Periodic Classification Of Elements

Q1.a

Q2.a

Q3.d

Q4.d

Q5.d

Q6.b

Q7.b

Q8.a

Q9.c

Q10.d

Q11.b

Q12.a

Q13.a

Q14.a

Q15.

a---c

b---d

c---b

d---a

Q16.

a---c

b---b

c---d

d---a